One to Five

Contents

1 One

When counting things you start at one.

There is one of me.
There is one of you.

Two

After one comes two.

Me and you make two.

3 Three

One, two, three. Two and one make three.

Dad, Mum and me
make three.

Four

Dogs have four legs —
one, two, three, four.

There are four of us.

Five

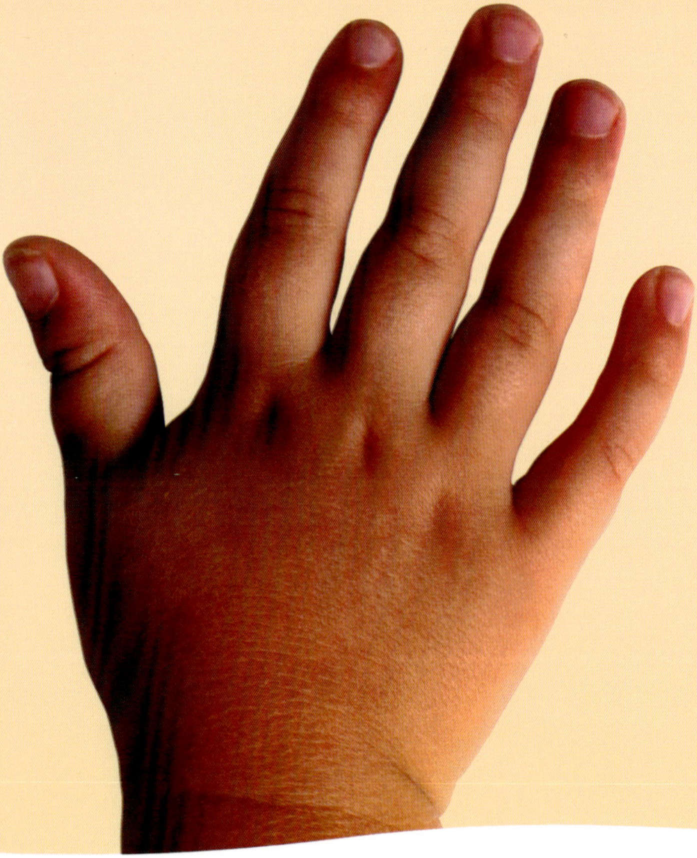

I have five fingers.

The five of us are friends.

One to Five

1	one		
2	two		
3	three		
4	four		
5	five